The Ethnology of the Pacific ... To which is added the discussion thereon. Reprinted from the Journal of Transactions of the Victoria Institute, etc. [With an ethnological map.]

Samuel James Whitmee

The Ethnology of the Pacific ... To which is added the discussion thereon. Reprinted from the Journal of Transactions of the Victoria Institute, etc. [With an ethnological map.]

Whitmee, Samuel James
British Library, Historical Print Editions
British Library
1879
20 p. ; 8°.
10492.cc.24.(1.)

The BiblioLife Network

This project was made possible in part by the BiblioLife Network (BLN), a project aimed at addressing some of the huge challenges facing book preservationists around the world. The BLN includes libraries, library networks, archives, subject matter experts, online communities and library service providers. We believe every book ever published should be available as a high-quality print reproduction; printed on- demand anywhere in the world. This insures the ongoing accessibility of the content and helps generate sustainable revenue for the libraries and organizations that work to preserve these important materials.

The following book is in the "public domain" and represents an authentic reproduction of the text as printed by the original publisher. While we have attempted to accurately maintain the integrity of the original work, there are sometimes problems with the original book or micro-film from which the books were digitized. This can result in minor errors in reproduction. Possible imperfections include missing and blurred pages, poor pictures, markings and other reproduction issues beyond our control. Because this work is culturally important, we have made it available as part of our commitment to protecting, preserving, and promoting the world's literature.

GUIDE TO FOLD-OUTS, MAPS and OVERSIZED IMAGES

In an online database, page images do not need to conform to the size restrictions found in a printed book. When converting these images back into a printed bound book, the page sizes are standardized in ways that maintain the detail of the original. For large images, such as fold-out maps, the original page image is split into two or more pages.

Guidelines used to determine the split of oversize pages:

• Some images are split vertically; large images require vertical and horizontal splits.
• For horizontal splits, the content is split left to right.
• For vertical splits, the content is split from top to bottom.
• For both vertical and horizontal splits, the image is processed from top left to bottom right.

THE

ETHNOLOGY OF THE PACIFIC.

By the Rev. S. J. WHITMEE,

A.V.I., F.R.G.S., ETC.

TO WHICH IS ADDED

THE DISCUSSION THEREON.

REPRINTED FROM THE JOURNAL OF TRANSACTIONS OF THE VICTORIA INSTITUTE, OR, PHILOSOPHICAL SOCIETY OF GREAT BRITAIN.

LONDON:

D. BOGUE, ST. MARTIN'S PLACE, TRAFALGAR SQUARE.
EDINBURGH: J. THIN.   DUBLIN: G. HERBERT.
PARIS: GALIGNANI & CO.

THE

E T H N O L O G Y

OF

T H E   P A C I F I C.

THERE are three classes of people inhabiting those islands of the Pacific Ocean which I include under the term Polynesia.   In the western islands, from the east end of New Guinea and Australia, eastward as far as and including Fiji, we find a black, frizzly-haired people; in all the eastern islands there are large brown straight-haired people (these are found also in New Zealand); and in the western islands north of the Equator there are smaller brown straight-haired people.

These three classes of people are represented in the map by the colours blue, pink, and purple respectively.

In a paper recently read before the Anthropological Institute, I have proposed the following names for these people.

To extend to all the blacks of the western islands the name *Papuan*,* which has long been applied to the black people of New Guinea, and some other portions of the Indian Archipelago. They have already been called Papuans by some writers, but are generally known as Melanesians. I believe these people are essentially like the Papuans of the Indian Archipelago, and that one name may serve for both.

For the natives of Eastern Polynesia and New Zealand I proposed an entirely new name, because there is no good general term by which they are known. This is *Sawaióri*.† For those living on the north-western islands I also proposed a new name, viz., *Tárapon*.‡ The adoption of these names may be objected to by some ethnologists; but as my reasons for proposing them have been given in the paper mentioned, which will shortly be published, I shall not defend them here. In the present paper I shall use these names, giving, however, the others by which the people have been hitherto known.

I. THE PAPUANS.

Melanesians, Negritos, Negrito-Polynesians, and Black Polynesians.

In colour these islanders are mostly black, or nearly so; but not of a jet black. Some are much lighter than others. It was long popularly supposed that their hair grew in small tufts. This was, however, a mistake which probably arose from the manner in which many of them are accustomed to dress it. On some islands the men collect it into small bunches, and carefully bind each bunch round with fine vegetable fibre from the roots up to within about two inches of the ends. Dr. Turner, in his "*Nineteen Years in Polynesia*," gives a good description of this process.§ He counted the number of bunches on the head of a young man, and found nearly seven hundred. He also calls attention to the strange

---

* "*Papuah*, frizzled, woolly-headed, having many natural curls."— Marsden's *Malay Dictionary*. "*Papuwah*, frizzled; the island of New Guinea; an inhabitant of that island, being of the Negrito race."—Crawfurd's *Malay Dictionary*.

† From *Sa*-moa, Ha-*wai*-'i, and Ma-*ori*, the names of three representative peoples belonging to the race.

‡ From *Tara*-wa, and *Pon*-ape, names of two representative islands in the Gilbert and Caroline groups respectively.

§ Pp. 77 and 78. Opposite to p. 76 is the figure of a Tanna man, which may be compared with the sculpture on p. 78.

resemblance existing between the hair of these people thus
dressed and the conventional representation of hair in the Assyrian sculptures, with which we are all familiar. When allowed
to grow naturally, the hair of the Papuans is always frizzly.

In the features of these people there is considerable difference. In a typical specimen the lips are somewhat
thick, the nose is broad, often arched and high, but coarse.
Their jaws project, and they may, as a rule, be said to be prog-
nathous. They are generally small in stature; but in some
islands the natives are comparatively large. Where, however,
they are of large size, we almost always have other evidence
of their mixture with another race. Speaking, therefore, of
the typical Papuans, we may say they are small, with thin
limbs, and are physically weak. In their natural condition
they are a savage, bloodthirsty race: one of the most savage
races of men living. They are invariably cannibals. As far
as I know, we have never yet come to know any portion of
the race without finding them addicted to this horrid custom.
They are also always broken up into small hostile tribes,
holding no intercourse with one another, except by warfare.
This is one of the most constant characteristics of the race.
The different languages spoken by them are very numerous,
owing, no doubt, to their hostility towards one another. A
missionary may learn the language of a tribe living in a par-
ticular valley, and on gaining access to a tribe in the next
valley, only a few miles distant, may find himself unable to
communicate with the people, owing to their language being
so different from the one he has learnt. In the grammatical
structure of these languages there is a considerable resemblance, as would naturally be expected; but owing to long
isolation through the savage disposition and hostility of the
people, the verbal differences have become very great.

Among them women hold a very low position. They are
merely the slaves and tools of the men. Nearly all the hard
work falls to their share, the men devoting themselves chiefly
to warfare. The women work at the plantations, carry the
burdens, wait on the men, and take their food from the leavings
of the lords of creation. The men will think they do well if,
with their arms, they protect the women from the attacks of
other tribes.

You may well imagine such a people to be in every respect
low in the scale of humanity. They are low socially, as we
have seen. Their family life is not greatly elevated above the
relationships existing among the lower animals. The relations
between the sexes are of the most degraded character, with a
few redeeming qualities. Affection is no doubt manifested

towards children, but even this is seen among the lower animals, and does not, of itself, indicate much tenderness of disposition.

In their mode of government might is right; and might is nearly the only thing which commands anything like respect. Intellectually too I should say the Papuans are low. As a rule they appear to lack the elaborate traditions and poems and songs found among many barbarous races. I think there are few indications among them of much power of mind. Religiously, too, they are low. They are not naturally a people possessing much religious feeling or reverence. Their religious systems, such as they have, are more of the nature of fetishes than anything else.

In arts and manufactures they are comparatively low, although there are some exceptions. Usually their houses are poor structures. On many islands the people have no boats, or their canoes are of very inferior construction. As a race they are not navigators. Their arms are, however, somewhat elaborately made, and most of them make a coarse kind of pottery. In some parts of the Solomon group the people build much better houses than those usually found among the Papuans; they also carve some of the woodwork in their houses rather elaborately, and build good sea-going boats. These things are, however, so exceptional that I am convinced they indicate contact and mixture with another race. In Fiji the people build good houses and good boats; but we know the Fijians are mixed with Sawaiori blood. I think it a justifiable inference that the Solomon islanders are also considerably mixed: and the reports of visitors to the group respecting the size, colour, and appearance of some of the people prove this inference to be correct.

Indeed throughout the whole of the Papuan region there is evidence of more or less mixture of the people with Sawaiori blood. In some islands there are pure colonies of the latter people, who keep themselves distinct from their blacker neighbours; but in many other places they have mixed with the black aboriginal population, and have considerably modified it. The map shows by pink patches and bands the positions of these colonies, and the extent of the mixture of blood as far as our present information goes.

Missionaries have ever found the Papuan race a difficult one to influence and elevate. They are lower and more savage than the Sawaiori people. There is less original capacity for the appreciation of that which is noble and good than we find among the others. There is in them less inherent religious feeling, or less of what may be called the religious instinct. There has therefore been more difficulty in finding a fulcrum upon which to rest the lever by which they are to be lifted. They no doubt, like other men everywhere, possess the capacity for religious belief and worship, but it is naturally of a low order. Hence Christian missions have been more difficult, and the success achieved has been less, in proportion to the means used, than among the other people of the Pacific.

The following broad characteristics of the Papuan languages I give in substance from a paper of my own recently contributed to the Philological Society. Consonants are freely used, some of the consonantal sounds being difficult to represent by Roman characters. Many of the syllables are closed. There is no difference between the definite and the indefinite article, except, perhaps, in Fiji. Nouns are curiously divided into two classes, one of which takes a pronominal postfix, the other which never takes such a postfix. The principle of this division appears to be a near or more remote connection between the possessor and the thing possessed. Those things which are connected with a person, as the parts of his body, &c., take the pronominal postfix. A thing possessed merely for use would not take this postfix. For example, in Fijian the word *luve* means either a son or a daughter—one's child, and it takes the possessive pronoun postfixed, as *luvena;* but the word *ngone*, a child, but not necessarily one's own child, takes the possessive pronoun before it, as *nona ngone*, his child, *i. e.*, his to look after or bring up.* Gender is only sexual. Many words are used indiscriminately, as nouns, adjectives, or verbs, without change; but sometimes a noun is indicated by its termination. In most of the languages there are no changes in nouns to form the plural, but a numeral indicates number. Case is shown by particles, which precede the nouns. Adjectives follow their substantives. Pronouns are numerous, and the personal pronoun includes four numbers — singular, dual, trinal, and general plural; also inclusive and exclusive. Almost any word may be made into a verb by using with it the verbal particles. The differences in these particles in the various languages are very great. In the verbs there are causative, intensive or frequentative, and reciprocal forms.

I have already said I believe these people belong to the same race as the Papuans of New Guinea and some other parts of the Indian Archipelago. Those who know the latter

---

* Hazlewood's *Fijian Grammar*, pp. 8 and 9.

people will recognize the characteristics which I have given as being almost equally applicable to both. It is for this reason I have proposed to use one name for all—the Eastern and Western Papuans, those of Polynesia and those of the Indian Archipelago.

In a lecture delivered last year at the Royal Institution Prof. Flower, F.R.S., virtually admitted that these people are alike, although he used different names for them. After speaking of those in Polynesia under the usual name Melanesians, he says :—

" People having very much the same physical characters as the Melanesians inhabit the islands of the Louisiade Archipelago, those of Torres Straits, and a very considerable part of New Guinea, and even some of the islands farther west, as Aru, Timor, Gilolo, &c. The exploration of New Guinea in an ethnological sense is only now commencing, and promises a most interesting feature. The greater part of the island is certainly inhabited by a dark-skinned race, with crisp or frizzled hair; indeed, the name by which they are frequently known, *Papuans*, is said to allude in the Malay language to the latter peculiarity. It is, however, very doubtful whether they all possess the uniform characters of the genuine Melanesian."[*] The last sentence refers to a now well-known mixture of races in parts of New Guinea, which I shall have occasion again to mention.

Until recently I should have said the eminent naturalist, Mr. A. R. Wallace, controverts the opinion of the essential unity of the Papuans of Polynesia and those of the Indian Archipelago. But his article in the *Contemporary Review* for February last[†] shows clearly that he has changed his view on this point since he wrote his "Malay Archipelago." In this recent article he speaks of "the Papuan or Melanesian." And in his description of the people he speaks indiscriminately of natives of New Guinea, New Caledonia, and the New Hebrides; and he gives the area occupied by the people as one " of which New Guinea is the centre, extending westward as far as Flores, and eastward to the Fijis."

I feel some satisfaction in noting this change in Mr. Wallace's views, because not many years ago[‡] I tried to prove he was wrong in believing that all the people of Polynesia belonged to one race, and had no relationship with the

---

[*] *Royal Institution Lecture*, May 31, 1878, p. 38.
[†] *New Guinea and its Inhabitants.* See pp. 426–428.
[‡] *Contemporary Review*, Feb., 1873.

inhabitants of the Indian Archipelago, and this is, I believe, the first time he has published a different opinion.

As to the wider affinities of these Papuans with other peoples of the world, I wish to speak cautiously. But I believe they may be remotely classed (together with all the other black people of the Southern hemisphere) with the tribes of Africa. In all essential respects they appear to be remote relatives, and the differences between them may probably be accounted for by (1) long isolation; (2) dwelling under different conditions in their various localities; (3) in some cases more or less mixture with other races.

Prof. Flower thinks the resemblance between the Papuans and African Negroes "which appears to strike every one who sees them for the first time, is rather superficial, and depending much upon colour and the character of the hair."* But Mr. Wallace, in the recent article I have already mentioned, says, "it is impossible not to look upon" these "Eastern Negroes" and the Africans "as being really related to each other, and as representing an early variation, if not the primitive type of mankind, which once spread widely over all the tropical portions of the Eastern hemisphere."† In the main, I take Mr. Wallace's view.

That the Papuans were the earliest occupants of the various places where remnants of the race are now found, and that they have, in many places, been partly or wholly overrun and displaced by more recent races, I think is unquestionable.

II. The Sawaiori Race.

Polynesians, Brown Polynesians, Malayo-Polynesians, and Mahoris.

These people are a large-sized race, their average height being about 5 feet 10 inches. They are well-developed in proportion to height. Their colour is a brown; lighter or darker, generally, according to the amount of their exposure to the sun; being darker on some of the atolls were the people spend much time in fishing, and among fishermen on the volcanic islands; and lighter among women, chiefs, and others less exposed than the bulk of the people. Their hair is black and straight, but wavy or with a tendency to curl in individual examples. They have very little beard. Their

---

* *Royal Institution Lecture*, pp. 37 and 38.
† *Contemporary Review*, Feb., 1879, p. 427.

features are generally fairly regular, eyes in colour invariably dark, and in some persons a little oblique. Jaws not projecting except in a few instances; lips of medium thickness, thicker than our own. Noses generally short, but rather wide at the bases. Their foreheads are fairly high, but rather narrow.

When young many of the Sawaiori people of both sexes may be spoken of as being fairly good-looking. The men, I should say, have more regular features than the women. The women, even if they are good-looking when they are young, soon lose their beauty. More attention was paid to personal appearance among the men than among the women; but such is not the case now.

As an uncivilized race the Sawaiori people are remarkable for their superior manners. They are a very polite people, and are far above mere savages. Indeed, there are many indications that they have descended from a state in some respects superior to that in which they were found at the time they were discovered by Europeans. The position occupied by women is one of these. Among this race generally women occupy a position hardly inferior to that of the men. Among the most polite and superior of the people women have as much influence and are treated with as much respect as among civilized races. They, in some instances, take hereditary titles and offices. It is well known to you all that a queen long reigned in Tahiti—Queen Pomare; and this is not an exceptional circumstance. Another indication of the comparative elevation of the people is the existence of rank and titles which are hereditary. Among most of these people as much is thought of rank as among ourselves. And so much deference is paid to chiefs that a different language is used in addressing them from that used to common people. Every part of a chief's body and all his belongings have different names from those appropriated to people of no rank. If a chief possess a dog the animal must be spoken of by a different name from that given to a common man's dog. The grade of rank of a person is indicated by some words addressed to him, three or four grades being recognized, and as many different terms being employed. For example, in Samoa there are four different words for to come, appropriated to four grades of people:— *sau,* for a common man; *maliu mai,* for a person of respectability; *susu mai,* for a titled chief; and *afio mai,* for a member of the royal family. When addressing a person in respectful language, the Samoans never use the first personal pronoun in the singular number, but always in the dual—the dual of dignity. This excessive politeness is sometimes somewhat puzzling or amusing to those newly arrived in the islands, and who may not have become accustomed to it. I remember the first time I noticed it I was riding a horse, and being met by a native he asked—" Where are you two going ?" a very ordinary mode of salutation to a person when on a journey. I at first thought he meant the question for me and my horse. But it was simply the dual of respect.

The way in which landed property is held and transmitted among the people also indicates something above savagery. It is not unlike the tenure of such property by the Israelites under the Mosaic laws. All the land in the islands is divided amongst families. An individual does not own it; but the members of the entire family have an equal right to its use; the patriarch or recognized head of the family, however, alone properly exercising the right to dispose of it, or to assign the use of it for a time to persons outside that family. Thus the land is handed down through successive generations under the nominal control of the recognised head of the family or clan for the time being. I use the word clan here, because the word family, in our sense of the term, does not express its full meaning among these people. A family is not the husband and wife and their children; but a whole clan, consisting of all the connections by blood and marriage. Each family or clan has a name, which is always borne by one of the oldest or most influential members, and the man who bears that name is the patriarch or head of the entire clan.

During the past few years this custom has been considerably changed in Samoa, and some of the larger families are broken into several sections—the nominal head of each section bearing the family title with a second name for the sake of distinction. In this way the binominal system is growing.

I believe these people once occupied a higher intellectual position than that they now occupy. They have most elaborate myths and songs—some poems being of considerable length, and I think superior in composition to anything the people were capable of at the time when Europeans first came into contact with them. The best of these songs and traditions are kept in Samoa in two forms—in prose and in poetry; and certain families are the recognised keepers of them. They were retained with great accuracy without being written—a father paying the greatest attention to teach them with verbal accuracy to his sons. The prose form of an important myth was not considered authoritative unless it agreed with the poetic form.

All the Sawaiori people were navigators before they were discovered by Europeans. Their boats were somewhat elaborately made and were very large. In them they made long voyages between the different groups. They sailed at certain seasons regulated by the appearance of certain constellations, and directed their courses by the stars. There can be no doubt but that a considerable amount of intercourse was kept up between the people in some of the distant groups in this way.

I think most members of the Institute will agree with me that all these characteristics taken together indicate that these people occupied a comparatively high level; and whether I have convinced you or not, I am myself quite satisfied that at the time of their first contact with Europeans there were indications that they had previously occupied a still higher position.

Let me now give you a few more of their general characteristics. As a race they are somewhat apathetic—differing, however, in different islands according to their surrounding circumstances. They live in an enervating climate, and on most of the islands nature is very lavish of her gifts. So they lead easy lives which foster an apathetic disposition. On the more barren islands and those more distant from the equator the people have more energy of character. All the people of the race think very well of themselves, and of some, at least, I should say they are very conceited.

As a people they are religiously inclined. They were strict and superstitious in their religious observances when they were heathen. Of them generally it may also be said, they were easily influenced by Christianity. They presented a contrast in this respect to the Papuan race. They possess a good measure of natural politeness—and in this respect the common people generally are immensely superior to the peasantry in our own country. I never met with a comparatively uneducated people who possessed more good common sense, and who would generally take a more reasonable view of things than the Sawaiori people with whom I came into contact. In every respect I may say they are a rather superior people.

The following brief sketch of the most prominent characteristics of their language may suffice for this paper. The phonology is simple. With one exception all the sounds found in them may be expressed by the Roman letters with their ordinary values. This exception is a sound which we call a "break," a kind of pause in the breath, which is between an aspirate and a *k*. A *k* sound takes its place in some of the languages. In those languages in which this sound occurs we usually write it by an inverted comma, as in the name *Hawai'i*. The vowel-sounds are all simple, as in Spanish. Every syllable is open. To this there is no exception. Some words consist entirely of vowels. Phonetic changes have taken place according to law, so that a given word in one language may have its form in any other language, if it be found in it, predicated. As a rule the accent is on the penultimate syllable; but in a few cases (chiefly when the last syllable ends in a diphthong or a long vowel, which is really a double vowel) on the ultimate. Very rarely, in some languages, the accent may be on the antepenult. There is an indefinite as well as a definite, and in some languages a plural article. Many words may be used as nouns, adjectives, verbs, or adverbs, without any changes of form. But some nouns are formed from the verb by taking a suffix, and some adjectives are formed from the noun in the same way. Gender is only sexual. There is some variety in the way of indicating number in the noun. In Samoa many nouns have special plural forms. The cases are indicated by prepositions. Proper names in the nominative case take a prefix, as *O Tahiti, O Samoa,* &c. Adjectives follow the substantives. The pronouns are numerous. Personal pronouns are singular, dual, and plural. The form of the plural in some languages shows that it was originally a trinal. In the verbs the distinctions of tense, mood, and voice are indicated by particles prefixed or postfixed. Number and person are generally regarded as accidents of the subject, and not of the verb. To this, however, the Samoan forms an exception; in this language many of the verbs have a special plural form. In all the languages there is a causative which is formed by a prefix to the verb. There are also intensive or frequentative, and reciprocal forms of the verbs. The intensive is usually a reduplication of the active verb; the reciprocal is usually formed by both a prefix and a postfix. Verbal directive particles are freely used, to direct towards, away from, or aside. In some languages, especially that of Samoa (I have already given examples above), many ceremonious words are used to persons of rank. Words which form part of the name of a chief are often disused during his life; and in some places they are disused after his death.

These languages are fairly copious, considering that they have been isolated and used by a people in small islands; and that until lately they have had no opportunity of gaining accretions from the outer world. Of the affinity of the people with other races, and the relationship which their languages bear to others, I will speak after describing the next people.

The Sawaiori race is, I think, very pure. In a few places it is, doubtless, a little mixed with Papuan blood; but this is only to a small extent. The people consider themselves superior to the black race; and while the black men will have brown wives, where the two races come into contact, whenever they can get them, I think a Sawaiori man would hardly have a Papuan wife, unless he could not get one of his own race. The Sawaioris occupy all the eastern islands in Polynesia from the Ellice group to Easter Island. There are also colonies of them found among the Papuans in the western area in the Loyalty Islands, the New Hebrides, and the Solomon group; and we now know that many of the inhabitants of the eastern portion of New Guinea resemble the Sawaiori people of Polynesia so much that they will most likely have to be classified with them. It is, however, probable that those on New Guinea are somewhat mixed. In the map I have indicated by pink bands in the Papuan area the relative proportion of Sawaiori mixture amongst the black race.

III. The Tarapon Race, or *Micronesians*.

In the western portion of Polynesia, north of the equator, there is a wide belt of low atolls or lagoon islands, usually known as Micronesia. Nearly all these atolls are peopled by a brown race of men in colour resembling the Sawaioris, but of smaller stature and less robust than they are. They have straight black hair, generally more lank than the hair of the Sawaiori people. These Tarapons, however, differ more among one another than the Sawaioris do. The natives of the Caroline Islands are, as far as I have seen them, much larger than those of the Gilbert group. They are also yellower in colour—more yellow than the Sawaioris, while the Gilbert Islanders are darker than the latter people.

I think there can be little doubt but these Tarapons are a people who are considerably mixed, and hence the differences which characterize them. In many respects they resemble the brown people of the Malay archipelago more nearly than they do the Sawaiori race. In fact, I think the bulk of the Tarapon people are the descendants of people who, in comparatively recent times, migrated from some portion of the Indian Archipelago; and that, since they have been living in those northwestern islands of Polynesia, they have become mixed with people of other races. Owing to this mixture, I always feel a difficulty in giving a general description which will apply to all the people in this region. The natives of the Carolines are, as I have already said, lighter than most of the others, and they differ in other respects, being larger than the Gilbert islanders, and less savage and warlike.

All the Tarapon people are navigators, and many of them build large boats, or proahs, not greatly unlike those found in the Indian Archipelago. Their houses are inferior to those of the Sawaioris. The arms of some are fairly well made, and in one group—the Gilbert Islands—they manufacture very elaborate armour to cover the entire body out of the fibre of the cocoa-nut husk. A corselet, which forms part of this, is a very ingenious and very good piece of workmanship, in shape not greatly unlike a piece of European mediæval armour.

Amongst them women appear to occupy a position not very different from that they hold among the Sawaioris, but somewhat lower. This difference is not in the amount of work and drudgery that they are expected to do, but rather in the social and domestic influence they exert. Religiously they are rather strict in the observance of their rites, and the shrines of their gods are very numerous. I visited some of the Gilbert Islands before any Christian influences had been brought to bear upon the natives, and in every house I saw a domestic shrine at which offerings of food, &c., were presented. In addition to these there were numerous other shrines in all parts of the islands.

Their gods were chiefly the spirits of their ancestors; the priesthood and chieftainship were combined in the same persons; they embalmed some of their dead, especially the bodies of beloved children; and women often carried about the skulls of deceased children hung by a cord around the neck as a token of affection.

The traditions of the Tarapons appear to be numerous. In some respects they resemble those of the Sawaioris. These deal very largely with the origin of their islands and the people. From them we learn that part of the people came from the west, and that these were met by some from the east. Most of the descendants of those arriving from the east were, however, destroyed by the others, who were the more numerous. As far as we have materials for examination, craniometry also indicates that the natives of these islands are more mixed than either of the other Polynesian races. Professor Flower, in his Royal Institution Lecture already mentioned, expresses that opinion, thus confirming the opinion which I have formed from an examination of the physical characteristics of the people, and from their languages.

In these languages consonants are used more freely than in the Sawaiori languages. They have some consonantal sounds which are not found in the latter, such as *ch, dj* and *sh* which may perhaps be regarded as intermediate between the Sawaiori and Papuan, although not nearly as strong as in the latter.

Closed syllables are by no means rare. Occasionally doubled consonants are used, but there is a tendency to introduce a slight vowel sound between them. In all of these particulars there is an approximation towards the Papuan. Most words take the accent on the penult. In some languages there appears to be no true article. In the Gilbert Islands language we find the Sawaiori *te* used in place of both the definite and indefinite article. Gender is sexual only. Number in the noun is either gathered from the requirement of the sense, or is marked by pronominal words or numerals. Case is known by the position of the noun in the sentence, or by prepositions.

In the language of Ebon—one of the islands in the Marshall archipelago, nouns have the peculiarity which I mentioned as being characteristic of the Papuan languages; viz., those which indicate close relationship—as of a son to his father, or of the members of a person's body—take a pronominal postfix which gives them the appearance of inflections. I do not know of the existence of this peculiarity in any other Tarapon language; but would not make too much of negative evidence.

Many words may be indiscriminately used as nouns, adjectives, or verbs, without any change of form. In some languages the personal pronouns are singular, dual, and plural. In others there are no special dual forms, but the numeral for *two* is used to express the dual. In the Ebon language there are inclusive and exclusive forms of the personal pronouns which, as far as I have at present been able to ascertain, do not occur in the other Tarapon languages. The verbs usually have no inflections to express relations of voice, mood, tense, number or person, such distinctions being expressed by particles. In the Ebon language, however, the tenses are sometimes marked—but even in that, the simple form of the verb is frequently given. All have verbal directive particles. In Ponape—one of the Caroline Islands—many words of ceremony are used only to chiefs, exactly as they are used so largely in Samoa. The custom of tabooing words which occur in the names of chiefs is also found there.*

I come now to consider the affinities of the Tarapon people and also of the Sawaiori race with other portions of the human family.

Both peoples may, I believe, be traced to the Indian archi-

* Most of the above particulars respecting these languages, and also those respecting the Sawaiori languages, I have taken from my paper already mentioned.

pelago; but further I shall at present not attempt to trace them. They have affinities with the Malays and other brown people now living in the islands of the Archipelago. But I wish you to understand that I do not think they have sprung from the Malay race as we at present know it. Doubtless, the Sawaioris are now more nearly in the primitive condition of the ancestors of the whole family than the Malays. I believe that at an early period (not later than the commence-ment of the Christian era, but probably earlier) the ancestors of the Sawaioris, the Tarapons, the Malays, and also the Malagasy of Madagascar, dwelt together in the islands of the Indian Archipelago. From some cause or other—probably war—a portion of that people migrated eastward to Polynesia. Finding the islands in the west occupied by the black Papuan race, they went on until they reached some of the islands in central Polynesia—perhaps Samoa—and there they settled. From this point they have spread abroad to the distant eastern islands; some have gone north-east to the Hawai'ian Archi-pelago; some have gone south-west to New Zealand; and a few others, at various times, have gone westward into the Papuan area, and have either formed colonies there, or have mixed with the Papuan people and intermarried among them. Some have, also, in comparatively recent times, gone north-west and mixed with the Tarapon people who entered Polynesia much later than the Sawaioris.

These Sawaiories being isolated from contact with other people have retained their primitive manners with consider-able purity, losing no doubt a good deal of what they originally possessed, owing to this isolation and to their living in small communities and on small islands. The changes which have taken place in them since their settlement have probably nearly all been losses, for want of circumstances to call for the use of some of the knowledge or habits which they possessed. There would be little or no addition to their knowledge, or change of any kind in the shape of accretions. Change would probably be entirely in the way of loss.

But, the people being naturally very conservative, the dis-integrating process would go on very slowly. This is shown by the remarkable similarity existing between their customs, their knowledge, and their languages over the vast space which they occupy. Hence I consider that these Sawaiori people at the present day represent very fairly the condition of the pri-mitive race from which they sprung at the time when they migrated from the common home.

The only time-mark which I know as giving an indication of the period of this migration, is the absence of Sanscrit elements in their languages. I should therefore say it was in pre-Sanscrit times : that is, before that language reached and influenced the languages in the Indian Archipelago.

At a later period a second migration took place from the Archipelago, and moved westward across the Indian Ocean to Madagascar. This, we may conclude, was in post-Sanscrit times—after that language had to some extent influenced the language of the people—for there are a few Sanscrit elements in the Madagascar language.

Later still—I think considerably later—another migration took place from the Indian Archipelago and went eastward, settling on the north-west islands of Polynesia, commonly known as Micronesia. The bulk of these people probably came from the Philippines, or some other islands in the north-eastern portion of the archipelago. The few Papuan elements which now appear to be in the Tarapon people may have been in the original people before they migrated. But since they have been settled in these islands there has, I believe, been a considerable infusion of other blood among them.

Part of this has come from the Sawaiori race. The traditions of the people mention Samoa as the place whence some of their ancestors came ; and I think we have good reason for believing that there is truth in these. But I believe other blood has been infused by the arrival of Japanese and Chinese junks with their crews at the islands. We have well-authenticated examples of such junks being driven across the North Pacific ; and I think it is highly probable that some of these have reached the islands of Micronesia, and that their crews have settled among the original people. I have given some evidence on this point in a paper recently published in the *Journal of the Anthropological Institute.* I need not therefore repeat it here.

The present paper has not been written with any controversial object. It has not been prepared from a special point of view for the Victoria Institute as distinct from other scientific societies. But from all that it contains, members of the Institute will see that no special arguments can be derived from Polynesia against the unity of the human family ; for all the three races inhabiting those islands have affinities with peoples in other parts of the world.

April 14th, 1879.

The CHAIRMAN.—I hope I may be allowed to return the thanks of the meeting to Mr. Whitmee for his very interesting paper. If any present desire to offer remarks upon the paper, now is the time.

Mr. ENMORE JONES.—I should like to ask a question upon a subject that has occupied my attention. On the 13th page of the paper the author deals with the religious notions of the Tarapon people, and asserts that their gods were chiefly the spirits of their ancestors. I should be glad to know what reasons they give for this belief. Has he been able to get from the natives any information as to why their gods are the spirits of their ancestors?

Mr. WHITMEE.—When I was in the Gilbert Islands I made inquiries on this point, and I found that they spoke of some of their ancestors who had migrated from other portions of the Pacific; some of them great men in their history, and regarded them as their gods—that is to say, they worshipped the memory of those ancestors. I have no doubt at all, from what I know of their traditions, that those persons who have been great men in their former history have become deified in that way.

Mr. JONES.—Then it is a mere matter of memory or recollection of persons on their part, in the same way that we respect the late Duke of Wellington or any other great man, but not a notion that the spirits of their ancestors are gods?

Mr. WHITMEE.—No; they believe that the spirit exists after death. That belief is universal in those islands, and it was for this reason that the women carried about with them the skulls of their dead children, and that the people buried their dead in their houses—in the family house. I asked them the reason why they did this, and they said, "We do it so that we may be together." They believe in the continued existence of the spirit after death.

Rev. T. M. GORMAN.—I should like, if I may be permitted, to follow up the questions that have just been put to Mr. Whitmee. At the Hibbert lecture last Thursday a similar point arose. It was in reference to where the most ancient Egyptians are represented as paying these honours to the memory of their ancestors. I am speaking of the most ancient form of the religion of Egypt. Those ancient Egyptians are represented as having made offerings of various kinds—fruits and fowls, beef, wine and beer—to the memory of their ancestors, and I was exceedingly struck with the idea that the priesthood were united in the same persons, which brings us more or less to the patriarchal relationship as we find it stated in our Bible—also the embalming of their dead and the partition, mentioned in to-night's paper, between the spirit and the body. For my own part I think these things show, among the islanders referred to by Mr. Whitmee, a striking resemblance to the ancient rites and ceremonials observed by the ancient Egyptians. I should also like to ask a question as to the use of the letters *l* and *r*. Has Mr. Whitmee noticed whether those letters are the same?

Mr. WHITMEE.—Yes, they are the same. The letters *l* and *r* are not distinguishable in Samoa. In the Samoan words in which the letter *l* occurs

c

the *l* sound is the common one, but it becomes *r* before the vowel *i* (or *e*, as we pronounce it) ; but these two letters are constantly interchanged throughout the languages. The Rev. Mr. Moulton, who has lived in the Tonga Islands, where he was connected with the Wesleyan missions, will be able to give you some information respecting the people of those islands. He is accompanied by a native gentleman.

Rev. J. E. MOULTON.—I did not come here with the intention of speaking; but as I have been called upon, I may offer a few words. Any person who has had personal intercourse with the race I have been amongst will acknowledge the wonderfully accurate manner in which it has been described in the paper we have heard read this evening. With very few exceptions I think I may say that what we have just heard coincides with my own experience after long personal knowledge. With regard to the division of the people made by Mr. Whitmee, I am quite certain it will ultimately be accepted on all hands. Writing on ethnology and geography at home is a very different thing from going out to the places treated of and acquiring a personal acquaintance with the people. Here we have to rely on the imperfect accounts given by the old navigators, and confirmed, I might almost say,—but, at any rate reappearing in our modern books and periodicals. I have seen very late editions of some of those books published for the guidance, or rather the misguidance, of our captains and sailors, and I have found in them the same errors which have been exploded numbers of times, and the repetition of which has in some cases led to mischievous results. Those who provide geographies without a personal acquaintance with the places and people have to depend on those old books. I remember that two or three years ago, having to write a geographical work for a college, I was led to precisely the same division as Mr. Whitmee, having had personal experience of the races mentioned. I trust that his designation of the Sawaiori race may be accepted; it is the only name that can correctly be given to that people, who, I think, have a right to be consulted in the choice of the name by which they are to be designated. Now, I belong to Tonga ; but at the same time I may say that there we cheerfully make way for our father, or mother, Samoa. I was forgetting that I was speaking in the presence of a gentleman from New Zealand, and perhaps he, as a Maori representative, will dissent from that statement; otherwise I think we shall all agree. We certainly do not think "small beer" of ourselves, and although we make way for Samoa, we are not content to be known under other names. Sawaiori we must regard as a very prominent group in Polynesia, and the term Sawaiori appropriately groups all that series of islands under one head. I have had Tongan pupils under me, and I think I may say that the people are certainly a most superior people. I remember reading in one of the books of the old navigators an account written by a captain who went out to that part of the world in very early times, and he spoke of those people as some of the finest savages in the world ; and I may add that under the influence of Christianity they have not at all deteriorated. Considering their isolation and opportunities, I think they will bear comparison with any of the races of the world. Of course they have not 1800 years of civilization behind them, but the wonderful progress they have made in civilization confirms the opinion that they are a superior race. If language is to be taken as an index, they must be acknowledged as very superior. Our modest lecturer has stated that the languages of the Sawaiori race are "fairly copious." I may add that I am collecting words for a lexicon of the Tonga language, and I believe I have already obtained 10,000. How those words can have been retained in circulation all these years without any printed book to preserve them I cannot understand. It is true that they have a number of songs in which numbers of ancient traditions have been embalmed; but I think you will all acknowledge with astonishment the vast number of words that have been retained—an amount that goes far beyond any comparison with the vocabulary of the agricultural labourers of this country. With regard to what has been said about their belief in the existence of the spirits of their ancestors, I fancy that the word "ancestor" is somewhat misleading. In Tonga there are many traditions of past ages. They represent Tubal Cain and Noah as spirits, and you can scarcely call them their immediate ancestors. There are, however, a few of later times. If this is the meaning attached to the word "ancestors" by Mr. Whitmee, I agree with him. They have a belief in the immortality of the soul after death, and they say that the soul keeps hovering about not very far from this world. This is their universal belief, and any idea to the contrary never entered the brain of a single Tongan.

Mr. WHITMEE.—I was speaking of the Tarapon race, and they speak of those who peopled their islands and the leaders of their expeditions as being their gods. I referred only to the leaders of these expeditions and the great men in their past history (most of them having existed at periods very remote) as those who have been deified.

Rev. J. SHARP, M.A.—Do they have images of those ancestors?

Mr. WHITMEE.—No, sir; I saw a great many of what I considered to be their stone gods, and I wanted to know what were the ideas they associated with those stones, and I found that they regarded the places where they were simply as shrines. I said, when I saw one, "Is that a god?" and they replied "No; that is the place where the god lives"—their gods are spirits: the shrines are simply the places where the gods are supposed to dwell.

Mr. MOULTON.—Are not the images wrapped round with the native cloth?

Mr. WHITMEE.—Not in the cases I have referred to; they are in some cases.

Rev. J. FISHER, D.D.—I should say that the people referred to in the paper who are likely to interest us most are the Sawaioris and the Tarapons; still I am a good deal interested in the Papuans. On page 7 the paper says, in reference to these people, "Mr. Wallace says, 'it is impossible not to look upon' these 'eastern negroes' and the Africans 'as being really related to each other and as representing an early variation, if not the primitive type of mankind.'" Now, I do not very clearly understand this. I do not understand what the writer means by "primitive." If it only means an early race, I can understand it quite clearly and accept it; but if it means that that was the primitive race, I think it is very difficult to accept the statement. I do not think the author of the paper would say that in the condition in which the Papuans are they have the ability or capacity to elaborate or construct a language consisting of verbs, adjectives, pronouns, and so on. We are told that the Sawaiori people have sunk or fallen; but we are not told that the Papuans have sunk, although the fact is that they have sunk more than the Sawaioris. Neither do I think that they represent the primitive type of mankind. On the contrary I think the primitive type was a different stock altogether, and that as the people went off from the primitive race they degenerated, losing all connection with their ancestors. They did not lose their language, but they lost many things which they possessed at the outset. I should like to know what meaning the author of the paper attaches to the term " primitive type," and whether he supposes that the people of that "primitive type" were equal to the construction of such a language as that of which he has spoken?

Mr. WHITMEE.—I may say briefly that I only quote in the passage referred to Mr. Wallace's words, and I go on to say—" In the *main* I take Mr. Wallace's view"; but on that point I do not take his view.

Rev. J. BULLER.—Does the author of the paper intend to indicate the extent of Christianity in New Zealand by the map which is exhibited?

Mr. WHITMEE.—I do not touch New Zealand. My map was prepared chiefly to illustrate missionary addresses. In speaking on missionary matters my subject is Polynesia, and I do not mention New Zealand, not having made that country a special study.

Rev. J. BULLER.—I am obliged to the author of the paper for that answer; but as I happen to have had a long residence in New Zealand I should like to say a word or two. The Maories of New Zealand, who are a very important branch of the Sawaiori race, do most certainly believe in the perpetual existence of their departed relatives, although they do not offer worship to them. Many of them have the art of ventriloquism, and without intending it they do by means of that art impose on the people generally. With respect to the intellectual capacity of the Sawaiorian race, I may say that some years ago when I was voyaging from Sydney to Auckland with Captain Markham and other naval officers, while sitting at the saloon table one day, a question arose with respect to Tongatapu, and I heard Captain Markham say he had been to the college under the care of my friend Mr. Moulton, and he said—"You will hardly give me credence: I am astonished at the progress of those boys, not merely in mental arithmetic, but in the higher branches of geometry. I do not think I could have passed some of them myself." That, I think, is a good testimony to the quality of mind possessed by those natives for acquiring the higher branches of learning. I might, if there were time, add other cases.

Mr. MOULTON.—May I ask why Mr. Whitmee did not include New Zealand in Polynesia?

Mr. WHITMEE.—I use the term to include all the islands in the inter-tropical regions, and New Zealand being out of the tropics I did not include it. I think it would be more naturally included under Australasia.

Mr. BULLER.—New Zealand is generally considered conventionally to come within the term Australasia.

Mr. GORMAN.—May I ask, is there any likelihood of the myths and songs of the Sawaiorians being published?

Mr. WHITMEE.—I hope there is some possibility, or even probability, that some of them will be published. Some of our missionaries have given great attention to the collection of those myths, and I know that one or two of them have obtained large collections. I have recently been urging them to contribute their gleanings to the lately established Folk Lore Society, who would be glad to have them. A book on the comparative mythology of Polynesia is one of the wants of modern times.

Mr. GORMAN.—Are there any traditions of the Flood and the Fall of Man?

Mr. WHITMEE.—Yes, sir; numerous traditions of the Flood.

Mr. GORMAN.—And of the Fall of Man?

Mr. MOULTON.—There are traditions of the primeval innocence of man.

Mr. BULLER.—Sir George Grey published a very large volume of poems and legendary tales which he had received direct from the Maories. The book was published at a guinea. I do not remember the publisher's name. It was published some years ago.*

Mr. GORMAN.—Photographs might be given showing the manner in which the hair of the people is worn, and the resemblance to what is found on the Assyrian sculptures. These would be important matters in regard to the connection of these people with Egyptology. With regard to the letters *l* and *r*, this is another matter of interest. In the Egyptian those letters are interchangeable, there being one character for each. In Abyssinia some missionary writer noticed that the people do not pronounce the letter *l*, but always make it *r*. These are three facts of great interest.

Mr. WHITMEE.—I should doubt the value of the fact with regard to the letters *l* and *r*, because I think it a very likely thing for different races to confound those letters. It is not only in Polynesia and Egypt that the *l* and *r* are interchanged.

Mr. MOULTON.—On that point the evidence is very misleading. In Tonga the letter *l* is plain and distinct and never approaches to *r*, and my difficulty in teaching the students the English language is in regard to the letter *r*. In New Zealand it is not settled, I believe, whether it ought to be represented by *l* or *r*.

At this point Mr. D. Finau, a native of Tonga, was called upon by Mr. Moulton to give the meeting an illustration by articulating the letter *r* distinctly in the word "rode": he did so, but subsequently failed to give the sound of the same letter in the word "drew." He then recited the Lord's Prayer in the native language.

---

* My copy of this work is dated 1853, and was printed by Robert Stokes, Wellington, New Zealand, but does not bear a publisher's name. The *Transactions and Proceedings of the New Zealand Institute* contain several useful papers on the Maories.—S. J. W.

Mr. BULLER.—There is no sound of the letter *l* in the Maori language, but the letter *r* has a sound approaching that of *d* in some words; nor have we a sibilant in Maori. (Mr. Buller here repeated the Lord's Prayer in the Maori language.)

Mr. WHITMEE.—With regard to the sibilant, that only occurs in the Samoan and the Ellice Islands. Samoa is called by most of the other natives Hamoa, with or without the aspirate.

Mr. MOULTON.—We have the sibilant in Tonga.

Mr. WHITMEE.—Yes, that is a third example in which it occurs in a comparatively few words.

Mr. E. SEELEY.—I may say with reference to the missions, that it is represented that the Papuan race receives Christianity more slowly than the other races, and yet the natives of the Fiji Islands receive it. There was some delay at first, but they have thoroughly received it now.

Mr. WHITMEE.—Yes, that is so; but it should always be remembered respecting the Fijians that they are not pure Papuans.

Mr. SEELEY.—I should like to know whether they have the same peculiarity in the skin which is said to mark the negro race? With reference to the question of the extinction of these races, it is an idea held by many at the present day that the degraded races of the world die out as they receive European civilization—that they are unable to bear European civilization. I do not believe this, and I should like to know whether the same process of extinction is going on among these races, that has gone on among some others with whom Europeans have come in contact? Is not this extinguishing process the fault of the Europeans rather than of their civilization? It is the custom now-a-days to identify people by their custom of land tenure, and the land tenure custom of these people is that of Ireland; but I do not suppose them to be related to the Irish in any sense: and if I am right, the question of land tenure is not a very serious thing.

Mr. WHITMEE.—The introduction of Christianity is more difficult among the black people than among the brown races, and this is certainly the case in the New Hebrides. In the Loyalty Islands the people have become Christians more rapidly than in others. We had a large force to go in and Christianize the people, who are somewhat mixed there. The asserted general decay of the Polynesians is an interesting question, which I should have liked to have discussed to-night; but it would have made my paper too long. Mr. Wallace, in a book which I only received last Saturday, says that these people are dying out; but he takes no notice of the statistics which show that they are not dying out. Some years ago Professor Rolleston delivered an address to the British Association at its meeting in Bristol, and he then gave some facts he had received from missionaries with regard to these people, showing that they were not dying out all over the Pacific. At that time he wrote to me for some further information, and I collected statistics, from which I found that while in some of the islands the people were most decidedly dying out, in other islands the previous decrease had stopped, while in some it had turned, through the influence of Christianity, to an increase. Where these races are dying out, it is owing to the fact that so-called civilized men went among them before Christianity was introduced with its beneficial influences. The white man went with his vices and strong drinks before the morality and religion of the Gospel were carried to those people, and thus the seeds of destruction were sown in the constitutions of the natives. But since Christianity has been introduced it has improved and benefited the people. I have great hope that some of those people will be spared to occupy an important position among the nations of the world.

Mr. SHARP.—The Secretary has asked me to say a word or two; and I may add one or two points touching on South India. I have been a missionary there and know something of the languages. There are one or two peculiar features in those languages that resemble those we have heard of to-night. In every one of the instances given in the paper it is said that gender is sexual only. That is the case in the extreme south of India. With regard to the personal pronouns having forms that are inclusive and exclusive, that is the case in Telugu and in Tamil. The difficulty as to the letters l and r appears in the Tamil language, and the proper pronunciation of "Tamil" is "Tamir," the r sound at the end being very hard. In certain parts of the country, however, the people pronounce the letter some as l and some as r. With regard to the relations between d and r, in Telugu the hard d merges so much into r that in translating it into the Roman lettering it is often given as r. In Tamil if they put two consonants together they slip a slight vowel sound in between. In the South Indian languages there is no article at all. (In Telugu every syllable is open.) As regards the patriarchal hold of property in India the property of a family is held, not as the property of the individual, but the elder brother (say) manages it for the rest as his co-proprietors. The point of the paper most interesting to me, is the conclusion, and the cautious words of the author as to whence these people have come and their relation to South India and the islands of the Indian Archipelago. No doubt the Dravidian race have migrated to some of the islands—Sumatra and Java for instance; but I do not know that they have got further.

[Mr. R. W. DIBDEN here referred to a recent number of the Journal of the Royal Geographical Society, and gave some extracts bearing upon the question as to whether cannibalism existed in New Guinea.]

Professor GRIFFITHS.—I have been called upon to say a few words; but I am a mere recluse and must trust to those gentlemen who have seen the various parts of the world for my facts, and do the best I can to generalize. There have been a large number of facts brought before us to-night, and it will be my business after I have gone home to think over them and make the best I can of them. I am deeply obliged to the author of the paper, whom I have heard often, and I was much interested in some of his remarks that have tended to confirm antecedent statements, my admiration being deepened by the caution with which he has put forward his facts. He has given them not only as a philosopher, but as a conscientious Christian, anxious not to overstep, but to bring out the truth. I am exceedingly obliged to him.

The CHAIRMAN.—I should like to ask Mr. Whitmee whether he can tell us about the class to which these languages belong? Can he tell us whether the languages of this Archipelago have belonged to a class of which the Sanscrit is an instance—whether they are Aryan languages, or whether anything is known generally as to the source from which they come?

Mr. WHITMEE.—I have not ventured to carry my studies much into the languages of Asia and Europe so as to trace the connection. I have been studying the Polynesian languages, and I have been urged to carry on my studies into some of those that are better known to scholars, but I have always said, I think if I use my special knowledge in the elucidation of the languages which have engaged my attention and bring them before the scholars of Europe they will be able to show the connection. I should be inclined to classify these languages, as far as I can see, with the Dravidian languages of southern India. With regard to cannibalism and the remarks that have been made as to New Guinea, it is necessary that we should be told the exact point of observation in New Guinea, as the people are so mixed there that a remark made about one point may not be applicable to another. We need more information with regard to the people of New Guinea before we can generalize to any great extent. As to cannibalism I know that there are cannibals in New Guinea at the present time and also in the islands round about; but there are people in New Guinea who are not cannibals. There is no doubt that the remarks made on this point are correct as far as they go, and that there are some parts where the people are not cannibals; but there are other points where they are cannibals.

A MEMBER.—Was there any knowledge of a Supreme Spirit before Christianity was introduced?

Mr. WHITMEE.—There was one great god, Tangaloälangi, who was worshipped all over Polynesia, and I have often thought that the traditions that exist with regard to this god may be some remnant of former knowledge which was much greater than what they now have. The name of the god I have spoken of means "Tangaloa, who dwells in the Heavens."

Mr. GORMAN.—The statement made in the Hibbert Lecture is strikingly illustrated by what is stated in this paper. The Egyptians addressed themselves to the spirits of their ancestors, and finished off by saying that they were faithful to the great god.

Mr. MOULTON.—The god just mentioned is called "the carpenter," or "the maker," or "framer." He has two brothers, and a sister is also mentioned, who was remarkable for her beauty. These things seem to point to the family of Tubal Cain; but of course this is only a matter of individual opinion. "Maui" was undoubtedly Noah; and it is strange that one of the sons of Maui is marked out as having been of exceptional badness, and his deeds are notorious. In the evenings spent by the young chiefs they generally talk about the exploits of this wild son of Maui. With regard to the dying out of the people, I must join issue with the author of that paper. When Europeans first went to these islands, they had not sufficient knowledge of the language to enable them to judge accurately of the facts. It did seem, at first sight, from the traditions, that the island of Tonga had been more populous. As I got a knowledge of the language, I had grave doubts about the extent of the population, and I was soon convinced that the evidence pointed to the fact that the island had not been much more populous than it is at present. A great number of reasons might be given to show that it was never more populous than now; but figures are the best arguments. It is difficult to get a reliable census. A missionary took the census of a considerable portion of the group at an interval of twenty years, and the increase, although in a place where the mortality had for a time been considerably above the average, had been at the rate of 25 per cent. for the twenty years, or, speaking roughly, 1 per cent. per annum. As soon as these islands have had Europeans upon them, they have had to stand the in-rush of our diseases, and they had also our vices brought amongst them. Just before I went to that island, whooping-cough visited it and swept away a very great number of the population. Two years afterwards it visited it again, and again carried off a great number. Now the disease is acclimatized, and the people take very little notice of it, although here and there a weakly child or an old person will be carried off by it. Influenza is another complaint that carried off a great number of persons— about 500 in three months—on that small island; now it is every year less virulent, and the mortality from it is not exceptional. There was a similar state of things in Samoa.

Mr. WHITMEE.—Mr. Moulton has given us the state of things in Tonga. In the book of Mr. Wallace to which I have referred, we have the latest information, and he tells us that the people of Tonga are dying out; and with regard to Samoa, he says the people are at present estimated at between 30,000 and 60,000. A few years ago I published in the columns of *Nature* the latest census of the Samoan islands, and it was there stated at 34,000 and a few hundreds, but Mr. Wallace, in the present month gives it at between 30,000 and 60,000. The fact is, that the estimates of former times were much too high, and on this point Mr. Moulton has evidently misunderstood the view I hold. I believe as fully as he does that no dependence whatever can be placed on these estimates. The decrease in the population of Polynesia generally is not as great as is usually supposed; but those who have arrived at preconceived opinions on these matters stick to them, and will not accept the facts we offer.

The meeting was then adjourned.

The Victoria Institute,

or

Philosophical Society of Great Britain,

7, ADELPHI TERRACE, STRAND, LONDON, W.C.

Correspondence (including communications from intending Members or Associates, &c.) to be addressed to "The Secretary." (For Subscription see next page.

THE PRIMARY OBJECTS.

THIS SOCIETY has been founded for the purpose of promoting the following Objects, which will be admitted by all to be of high importance both to Religion and Science :—

First.—To investigate fully and impartially the most important questions of Philosophy and Science, but more especially those that bear upon the great truths revealed in Holy Scripture, with the view of reconciling any apparent discrepancies between Christianity and Science.

Second.—To associate MEN OF SCIENCE and AUTHORS* who have already been engaged in such investigations, and all others who may be interested in them, in order to strengthen their efforts by association, and by bringing together the results of such labours, after full discussion, in the printed Transactions of an Institution ; to give greater force and influence to proofs and arguments which might be little known, or even disregarded, if put forward merely by individuals.

[For the special advantages secured to Country and Colonial Members and Associates in the Journal of Transactions see below.]

Third.—To consider the mutual bearings of the various scientific conclusions arrived at in the several distinct branches into which Science is now divided, in order to get rid of contradictions and conflicting hypotheses, and thus promote the real advancement of true Science ; and to examine and discuss all supposed scientific results with reference to final causes, and the more comprehensive and fundamental principles of Philosophy proper, based upon faith in the existence of one Eternal God, who in His wisdom created all things very good.

The Journal of Transactions

Is arranged so as to secure its special usefulness to Country and Foreign Members and Associates (who form two-thirds of the Institute). It contains the Papers read at the Meetings, and the Discussions thereon.

Before they are published in the Journal, the papers themselves, and the discussions, are revised and corrected by their Authors, and MS. comments and supplementary remarks are added, which have been sent in by those Home and Foreign Members to whom, as being specially qualified to pronounce an opinion upon the respective subjects, proof copies of the Papers have been submitted for consideration. These arrangements, which cannot but add to the value of the Journal, are carried out with a view to the advantage of , especially Country and Foreign Members, who thus find in the Journal ach valuable matter, in addition to that which had come before those a ally present at the Meetings.

* The Society now consists of 750 Subscribing Members (NEARLY TWO-THIRDS OF WHOM UNTRY AND FOREIGN MEMBERS) ; including His Grace the Archbishop of Can- and other Prelates and leading Ministers of Religion, Professors of English and n Universities, Literary and Scientific Men in general, and others favourable to the

MEMBERSHIP AND SUBSCRIPTIONS.

Intending Members and Associates are requested to apply to the " Secretary."
The Annual Subscription for *Members* is *Two Guineas,* with *One Guinea*
Entrance Fee ; (see privileges).   The Annual Subscription for *Associates* is *One Guinea,* without Entrance Fee.

In lieu of Annual Subscription, the payment of *Twenty Guineas* (without Entrance Fee) will constitute a *Life Member,* or *Ten Guineas* a *Life Associate.*

The payment of a Donation of not less than *Sixty Guineas* qualifies for the office of *Vice-Patron,* with all the privileges of a *Life Member* or *Life Associate.*

[It is to be understood, that only such as are professedly Christians are entitled to become *Members.*]

*** All Subscriptions are payable to the " VICTORIA INSTITUTE'S " credit at Messrs. " Ransom," 1, Pall Mall East, S.W., or may be remitted to " W. N. WEST," Esq., (the Treasurer), at the Institute's Office.

PRIVILEGES.

MEMBERS—on election, are presented with any Volume of the First or Second Series of the *Journal of the Transactions,* and ARE ENTITLED—to a Copy of the Journal, either in the Quarterly Parts, or the Annual (bound) Volume, for the years during which they may subscribe, and to a copy of any other documents or books which may be published under the auspices of the Society in furtherance of Object VI., and, on application, to a copy of every paper published in the " People's Edition"; to the use of the Library (Books can be sent to the country), Reading and Writing Rooms (affording many of the conveniences of a Club) ; and to introduce two Visitors at each Meeting.   The Council are chosen from among the Members, who alone are eligible to vote by ballot in determining any question at a General Meeting.   Members are further privileged to obtain any Volumes, other than that chosen, of the Transactions *issued prior to their joining the Institute* at half-price (half-a-guinea each), or any Quarterly Parts for past years at half-a-crown each.

> The Library, Reading and Writing Rooms are open, for the use of the Members only, from ten till five (Saturdays till two).   The Institute exchanges Transactions with the Royal Society and many other leading English and Foreign Scientific bodies, whose transactions are therefore added to the Library.

ASSOCIATES—ARE ENTITLED, to the Journal, in Quarterly Parts or in the Annual Volume, for the years during which they may subscribe ; to obtain the earlier Volumes or Parts at a reduced price ; and to introduce one Visitor at each Meeting.

Members and Associates have the right to be present at all Meetings of the Society.

The Meetings, of which due notice is given, are held at 7, Adelphi Terrace, at Eight o'clock on the evenings of the First and Third Mondays of the Winter, Spring, and Summer Months.   Proof Copies of the Papers to be read can be had by those desirous of placing their opinions thereon before the Members (when unable to attend, they can do this in writing).

Members and Associates on 1st January, 1871, 203.   Joined since.—In 1871, 91;—1872, 109 ;—1873, 110 ;—1874, 111 ;—1875, 115 ;—1876, 107 ;—1877, 100.

Members and Associates joined during 1878.

Foreign and Colonial 30
Country 57
London (*residing within Postal District*) 14

THE

OGY OF THE PACIFIC.

THE REV. S. J. WHITMEE,

A.V.I., F.R.G.S., ETC.

TO WHICH IS ADDED

THE DISCUSSION THEREON.

M THE JOURNAL OF TRANSACTIONS OF THE VICTORIA
GE, PHILOSOPHICAL SOCIETY OF GREAT BRITAIN.

LONDON :

MARTIN'S PLACE, TRAFALGAR SQUARE.

GH : J. THIN.   DUBLIN : G. HERBERT.

PARIS : GALIGNANI & CO.

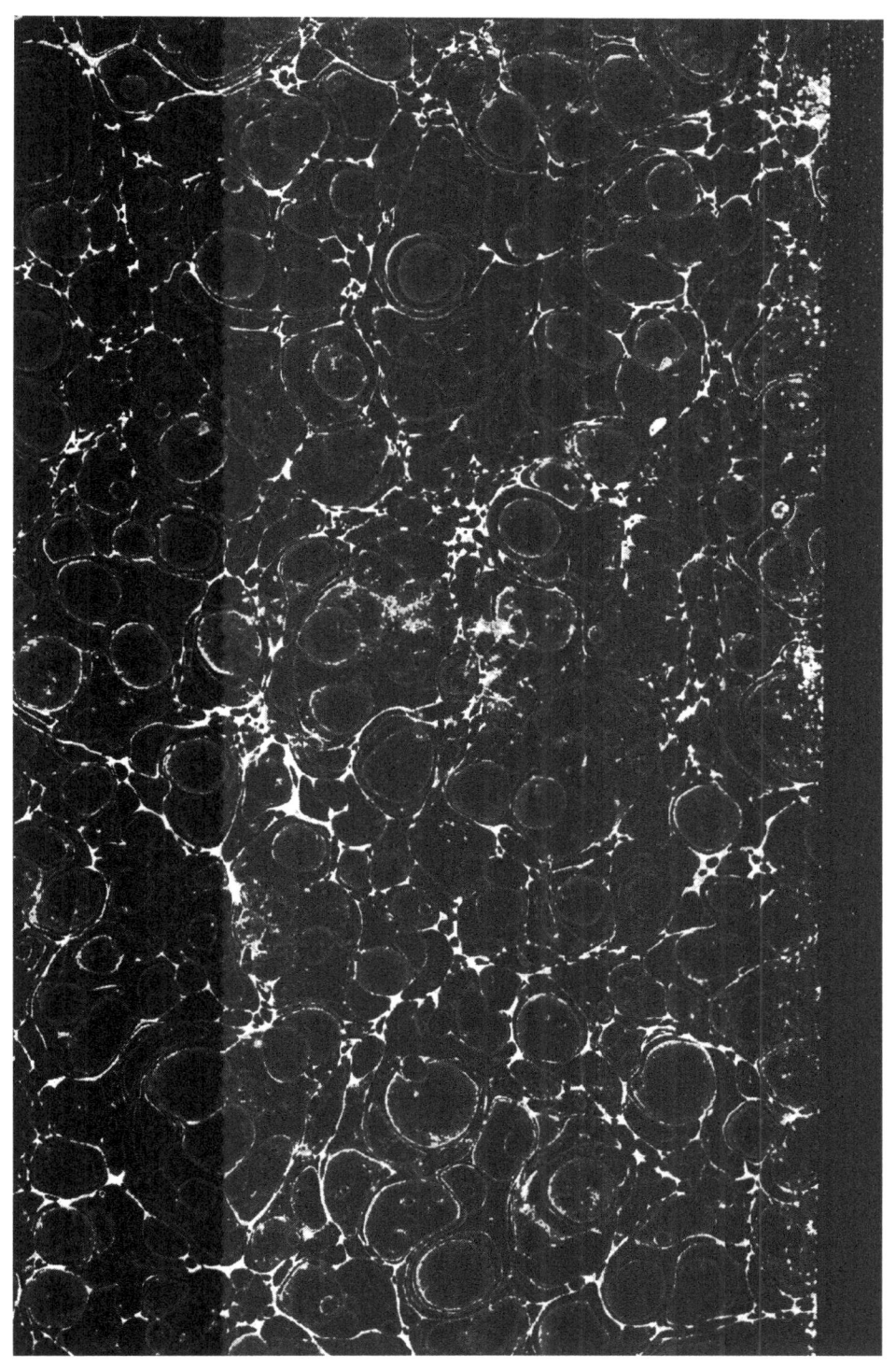

Lightning Source UK Ltd.
Milton Keynes UK
UKHW03f0619100818
327043UK00005B/230/P